Math Concept Reader

My Counting Trip to the Zoo

by Amy Ayers

Copyright © Gareth Stevens, Inc. All rights reserved.

Developed for Harcourt, Inc., by Gareth Stevens, Inc. This edition published by Harcourt, Inc., by agreement with Gareth Stevens, Inc. No part of this publication may be reproduced or transmitted in any form or by any means, electronic or mechanical, including photocopy, recording, or any information storage and retrieval system, without permission in writing from the copyright holder.

Requests for permission to make copies of any part of the work should be addressed to Permissions Department, Gareth Stevens, Inc., 330 West Olive Street, Suite 100, Milwaukee, Wisconsin 53212. Fax: 414-332-3567.

HARCOURT and the Harcourt Logo are trademarks of Harcourt, Inc., registered in the United States of America and/or other jurisdictions.

Printed in China

ISBN 13: 978-0-15-360164-4
ISBN 10: 0-15-360164-7

6 7 8 9 10 0940 16 15 14 13 12 11 10 09

Aunt Nina took me on a trip.
We went to the zoo.

The zoo is a fun place.
It is my favorite place to go!

The zoo has many animals.
I told Aunt Nina facts about them.

We will count the animals.
1, 2, 3, go!

First we saw 2 owls.
Owls hunt for food at night.

We saw 7 deer.
There were more deer than owls.

We saw 6 seals.

Seals can hear under water.

We saw 3 bears.

We saw fewer bears than seals.

We saw 8 turtles.

Turtles love to lie in the sun.

We saw 11 bats.
I know 11 is greater than 8.
There were more bats than turtles.

Aunt Nina and I got hungry.
We shared a picnic.

We ate near a pond.
Then we counted more animals.

We saw 14 ducks.
Ducks' feathers help keep them dry.

Next we saw 9 fish.
I know 9 is less than 14.
We saw fewer fish than ducks.

We saw 4 chimps.
Chimps love to eat bugs.

We saw 10 penguins.
I know 10 is greater than 4.
We see more penguins than chimps.

We saw 20 sheep.

Sheep know how to climb rocks.

We saw 18 storks.

I know 18 is less than 20.

We saw fewer storks than sheep.

Then we looked for snakes.
Snakes like to hide.
Snakes are good at hiding.

We saw 4 wolves last.
I know 4 is greater than 0.
We saw more wolves than snakes.

I took photos.
I will make a scrapbook.

I will share it with Aunt Nina.
It will help her remember our day!

Glossary

fewer There are fewer bears than wolves.

more There are more ducks than bats.

Photo credits: cover, p. 20 © Gary D. Landsman/Corbis; p. 2 © Nik Wheeler/Corbis; p. 3 © William Manning/Corbis; p. 4 © Lon C. Diehl/Photo Edit; p. 5 Adam Jones/Visuals Unlimited/Getty Images; p. 6 U.S. Fish and Wildlife Service; p. 7 Raymond Gehman/National Geographic/Getty Images; p. 8 © Wolfgang Kaehler/Corbis; pp. 9, 24 (top left) © Mark Newman/FLPA; p. 10© Jim Merli/Visuals Unlimited; pp. 11, 24 (bottom left) © David Hosking/FLPA; pp. 12, 23 Russell Pickering; p. 13 © Inga Spence/Visuals Unlimited; pp. 14, 24 (bottom right) © Klaus Hackenberg/zefa/Corbis; p. 15 © Warren Morgan/Corbis; p. 16 © Mary Kate Denny/Photo Edit; p. 17 © Galen Rowell/Corbis; p. 18 Nina Buesing/Stone+/Getty Images; p. 19 © Eric Wanders/Foto Natura/FLPA; p. 20 © Gary D. Landsman/Corbis; pp. 21, 24 (top right) Guy Edwardes/The Image Bank/Getty Images; p. 22 Jaimie D. Travis/DK Stock/Getty Images.

24